W9-CBU-188

FARM
Through the Ages

PHILIP STEELE

Illustrated by
ANDREW HOWAT and GORDON DAVIDSON

Troll Associates

Library of Congress Cataloging-in-Publication Data

Steele, Philip
Farm through the ages / by Philip Steele, illustrated by Andrew Howat and Gordon Davidson
p. cm
Summary: Surveys the history of farming and how it has changed both the lives of humans and the face of the landscape over thousands of years.
ISBN 0-8167-2731-7 (lib. bdg.) ISBN 0-8167-2732-5 (pbk.)
1. Agriculture–Europe–History–Juvenile literature.
2. Agriculture–History–Juvenile literature. [1. Agriculture–History.] I. Howat, Andrew, ill. II. Davidson, Gordon, ill
III. Title.
S452.S74 1993
630'.9–dc20 91-37819

Published by Troll Associates

Design by James Marks

Printed in the U.S.A.

10 9 8 7 6 5 4 3 2 1

Introduction

There are few wild places left in the world. Most of the countryside we see has been changed by humans. Today's woods were once part of vast forests. Today's fields of wheat were once a tangle of thorn and nettle. This book tells the story of farming. It tells how men, women, and children shaped the countryside over thousands of years, growing food and tending herds.

Our story is set somewhere in Western Europe. Farming did not begin in a single place or at a single time. It was practiced in the Middle East as early as 9000 B.C. But Africa, India, China, and the Americas all played their parts in the story.

Contents

Stone Age farmers

Sunshine and rain had sustained the valley for millions of years. The landscape had changed many times. For long periods, it had been a simple patchwork of white snowfields and dark green pine forests. In times of drought, it changed and became brown and parched.

By 3000 B.C. farming had reached our remote valley. A corner of the forest was burned and cleared. Wheat was grown there—a poor, straggly crop, but a welcome source of food. Crops were harvested with sickles and the grain was removed by beating. It was then shaken, so the husks blew away, and ground into a coarse flour.

Plants had spread far and wide, their seeds scattered by the wind. The soil of the valley was fertile and drained by a broad river. Each year, the spring floods left behind a deep layer of rich mud. The first humans took fish from the streams and marshes, and gathered berries and roots in the woods. They also hunted in the forest with flint-headed spears.

Far away to the east, New Stone Age people had learned how to harvest wild grasses, or cereals. They had learned how to sow and gather seeds. They had tamed sheep and goats to provide meat, milk, and clothing.

Plowing the land

New peoples with new ways settled in the valley. Some had learned the art of working copper and bronze. The Celts were skilled ironmasters, and by 200 B.C. they were using the hard new metal to make farm tools.

Cattle were kept for their milk. Some were slaughtered each autumn. Beef was salted down for the hard months of winter. Cattle herds were closely guarded against raids by warlike neighbors.

The fence enclosed ponies for riding. Sheep provided wool, which was spun into yarn and woven into cloth on simple upright looms. The cloth was dyed with lichens, berries, and flowers.

Barley and wheat were ground by hand in a quern, a hand-turned mill made of two stones. The grain was baked into bread, or mashed to make porridge that was sweetened with honey. Rye and oats were also grown as crops.

The soil was too heavy to turn with a wooden hoe. A team of oxen dragged a plow across the field, to break it up.

The Roman villa

Roman soldiers invaded the valley in A.D. 55, and built a camp near the old village of the Celts. Many of the tribesmen fled to the west. Others remained in the village. Roads were built across the countryside, and over the years the Roman camp became a small town.

When the Roman soldiers retired, some of them stayed on and built small farms near the town. They sold their produce at the market. Most of the countryside was still farmed in the traditional Celtic way.

In 250, a rich Roman merchant built a large country house, or villa, on the slopes of the valley. He had read all about farming methods and was eager to try out new ideas. Some of them failed. The vines he imported from Italy did not thrive in this cooler climate. He did try out a locally built reaper. A horse pushed a wheeled cart into the standing wheat. Iron teeth attached to the back of the cart pulled the ears of grain from the plant.

Hard work on the villa was done by laborers and slaves. They had to spread compost on the soil and dig drainage ditches with iron hand tools. They worked all day, until their limbs ached.

Strip farming

The Roman soldiers left the valley in 400 and never returned. The villa fell into a state of ruin. Thistles and weeds blew in the wind, where once there had been fields and orchards. The old village was burned to the ground by invading armies. There were many harsh years when the people starved.

In 800, a fine new village was built higher up the river valley, and a large area of forest was cleared. The timber houses were thatched with rushes. People planted rows of peas and beans. The river powered a water

The big fields were divided into strips. Each farmer had about 30 acres (12 hectares) of strip land and shared the ox team with his neighbors. Heavy wooden plows, with wheeled steering, cut the earth and made deep furrows.

mill, which was used to grind wheat into flour.

The village had three main fields. The crops planted in each were changed from one year to the next. Wheat was grown in one and oats in the other. The third was allowed to lie unplanted, or fallow. This ensured that the goodness of the soil was not used up too quickly. Cattle grazed freely on common ground and on the stalks, or stubble, left behind after the big fields were harvested.

Lords and peasants

The year 1050 saw more armies trampling across good farmland, stealing livestock, and burning fields. A Norman lord built a wooden castle to guard the valley, and over the years this became a great stone fortress.

The people of the valley had to serve the lord of the castle by providing farm labor or crops. Rents were high and taxes were cruel. When the lord built a new windmill nearby, he forced everybody in the valley to have their wheat ground there at a very high price.

The way in which the land was used hardly changed. The fields were still worked in strips, each the distance that an ox could plow without resting. Pigs rooted for acorns in the woods, guarded by a swineherd.

In the 13th century, large horses began to be used for pulling carts and plows. A new padded-collar harness was designed, which helped horses to pull heavy loads. Their hooves were now protected by iron horseshoes that were fitted by a blacksmith.

The people of the valley had their grain carefully weighed and checked by the lord's men to ensure that they did not have more than their share.

Wealth from sheep

The valley saw many changes. In 1348, a terrible plague swept across Asia and Europe. This "Black Death" killed two thirds of the farm workers in the valley.

There was a shortage of labor, and peasants could now demand to be paid a wage instead of giving their services to the lord of the castle. Some tenants bought their own farms and hired people to work for them.

The old landowners were unhappy with this new state of affairs. They introduced laws to keep wages low. The peasants were furious. They armed themselves and burned down the barns on the castle estates.

Many landowners turned to sheep farming, which employed fewer people and brought in good profits. The wool was sold to weavers, and the skin was sold to monks who made it into vellum, or parchment, on which they wrote out the Scriptures. Greedy landowners fenced in much of the common land around the village and used it to graze their own sheep.

The new estate

One fine day in 1650, a young cavalry officer rode up the valley. A new mansion had been built on high ground above the village. Laborers were digging a drainage ditch along the edge of a field.

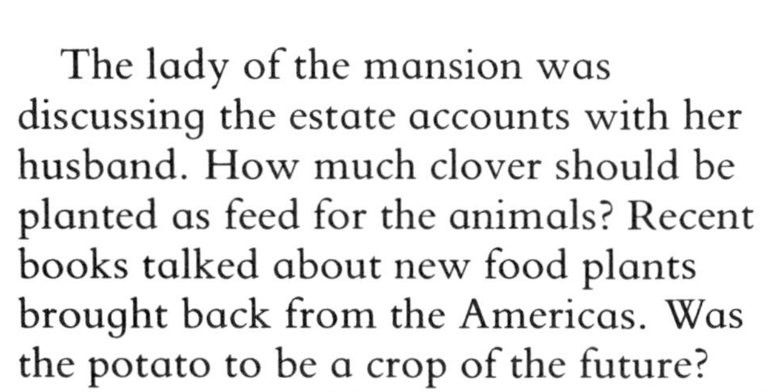

The lady of the mansion was discussing the estate accounts with her husband. How much clover should be planted as feed for the animals? Recent books talked about new food plants brought back from the Americas. Was the potato to be a crop of the future?

The cavalry officer rode into the yard bringing news of battles. The lady hoped that war would not return to the valley. What would become of their plans for the estate?

Many landowners raised birds. Pigeons and their eggs provided food. Geese were cooked for a large meal, and turkeys had been brought to Europe from North America. Wildfowl, such as duck and quail, were also eaten.

New farming methods

On the lands of the big estate, the talk in 1740 was of new farming methods. Plows were improved and built of lighter, stronger wood. Seeds were no longer scattered by hand, but sown with a horse-drawn seed drill. This was a wheeled box, from which the seeds dropped down tubes into the furrows.

Land was no longer left fallow. It was found that by regularly changing, or rotating, the kind of crop grown, the soil could be used more efficiently. Sheep were now turned out onto turnip fields and left to forage during the winter. This meant that there was no need to slaughter the livestock each autumn. Fresh meat could be eaten during the winter months.

The smaller farms in the valley could not afford to farm in the new way. Most of these farmers were tenants who rented their land. If they had tried to improve their fields, they would have been charged higher rents.

The milkmaid crossed the farmyard from the dairy. She had been up milking the cows at first light and her hands were cold and chapped. Warm milk slopped in the pails that hung from the shoulder frame, or yoke, across her shoulders.

Farming machinery

The valley looked well cared for. Its hedges were trimmed and its ditches well dug. Potatoes grew in the top fields.

But the peace of the countryside was shattered. A copper mine had been opened in 1810, and now a steam locomotive hauled cars of ore down the valley. The engine's huffing and puffing scared the horses working in the fields.

Machines were now used on the estate, too. There was a thresher, a reaper, and an iron plow. There were even rumors that it was possible to plow a field using steam power.

The new machines helped farmers provide grain and vegetables for the growing population of the cities. Drovers led herds of cattle across the country to the city markets.

Not everyone liked the new machinery. One machine could do the work of several people, and farm laborers were afraid of losing their jobs. One summer day in 1835, they stormed across the estate yard and began to smash reaping and threshing machines. The ringleaders were later caught and severely punished. Others left for the city to work in the new factories and textile mills.

The agricultural society

New machines were seen on the farms, but at harvest time every family in the village still toiled in the fields. Wheat was cut by hand and bound into bundles called sheaves. Scientists were now studying chemistry and biology in order to improve the soil and crops. New kinds of cattle, sheep, and pigs were produced over the years by selecting only the best ones for breeding. This had been done for hundreds of years, but now the farmers used scientific methods.

In 1870, the farmers of the valley founded an Agricultural Society. They met to discuss the latest articles in the farming papers, and held an annual show in the local market town. Prizes were given for the best bulls, sheep, and horses.

The farmer looked proudly at his prize bull as it was led into the market ring. Such a fine animal should fetch a fine price. Behind the cattle ring, a line of laborers, shepherds, and foremen shuffled nervously. They were waiting to see who would be hired by the farmers. They would be paid low wages for very hard work.

A new tractor

The winter of 1935 was a harsh one, and several lambs were lost in a blizzard. The new pipes that brought water from the village cracked and burst, and the farmer had to use the old water pump in the yard.

A fallen tree blocked the lane, and the farmer had to borrow a tractor

Tractors became more common during the 1930s. In the future, they would be used to pull plows and seed drills and to spread fertilizer.

from the big estate to haul the tree clear. The tractor's big metal wheels gripped the icy ground as it pulled the tree off the lane. Most hauling work in the valley was still done by horses.

Times were hard. Low-priced food imported from abroad could now be bought in the local shops, and the competition was driving farmers out of business. Many laborers went hungry. Some joined a farmers' labor union. They paid in a small sum of money from their wages each week, and the union campaigned to protect their jobs.

Farms or factories?

Farms were more efficient than ever before, and life in the valley was much easier.

During the 1960s some farms were beginning to look like factories. Chickens and turkeys were now raised in cages called batteries, housed in long sheds. Cows were given numbers instead of names, and electric milking machines had been installed in the cowsheds. The milk was collected by tanker each day and taken to dairies in the city.

Giant combine harvesters made short work of whole fields of wheat, cutting, threshing, and baling in one

operation. Excavators could dig ditches and remove hedges. Liquid fertilizers were sprayed on the soil, and crops were protected from pests by insecticides. Some of these chemicals poisoned wildlife and polluted local streams.

Fewer people worked on the land, but their pay and living conditions had improved. Some of the old farm cottages were bought by city dwellers or by people who drove to work in local towns each day.

The present

The farmer checked his accounts in the farm office. How much had he paid for the veterinarian to help with calving last year? It was all filed on the computer. Farming was big business these days and had changed so much since he was a boy.

It had been a hot summer. The crops would need plenty of water if the harvest was to be a success. If the harvest failed, the farmer might try different crops or even a different way of farming. There were now so many choices open to him.

In the yard, a young student from the agricultural college was fixing the spreader. She was training to be a farmer herself. Modern tractors were complicated, powerful machines with up to 24 gears.

Organic food, grown without the use of chemicals, was expensive to produce. But it was selling well. The farmer already produced broccoli and carrots this way. Modern science and technology had changed farming, but people still wanted vegetables that tasted home grown.

The future

The valley had been farmed for over 6,000 years. In the last 200 years it had seen more changes than in all those other centuries put together. What did the future hold?

A healthier, cleaner way of farming was needed. The valley had become polluted and poisoned. Animals had sometimes been treated cruelly as farmers tried to make greater profits. New farming methods had sometimes produced diseased livestock and unhealthy food. Some people said that pollution of the air by factories and motor vehicles in the cities might even change the climate.

Would new crops be bred that could survive heat and drought? Such crops were already needed in Africa and Asia. Europe, America, and Australia now produced huge amounts of food, while in other parts of the world people were starving. Farming needed to be organized worldwide.

During the 1980s, plants were grown in space for the first time on the space-shuttle missions. Perhaps in the future, farms would be built inside space stations and the harvest ferried back to Earth. Farming has come a long way since people first began planting seeds thousands of years ago. But we will always be looking for better ways to produce the food we need.

Index